# THE AMAZING WORLD OF FLYINGFISH

# The Amazing World of
# Flyingfish

Steve N. G. Howell

Princeton University Press

Princeton and Oxford

Copyright © 2014 by Steve N. G. Howell
Published by Princeton University Press, 41 William Street,
Princeton, New Jersey 08540
In the United Kingdom: Princeton University Press,
6 Oxford Street, Woodstock, Oxfordshire OX20 1TW
nathist.press.princeton.edu

Cover photograph: Ornate Goldwing in the Western Tropical Pacific, April 10, 2008.
© Steve N. G. Howell

Library of Congress Cataloging-in-Publication Data
Howell, Steve N. G.
The amazing world of flyingfish / Steve N. G. Howell.
pages cm
Includes bibliographical references and index.
ISBN 978-0-691-16011-5 (hardcover : alk. paper) 1. Flyingfishes.
I. Title.
QL638.E9H69 2014
597′.66—dc23
2013039955

British Library Cataloging-in-Publication Data is available
This book has been composed in Minion Pro and Scala Sans OT
Printed on acid-free paper.
Printed in the China
10  9  8  7  6  5  4  3  2  1

Dedicated to the extended Howell clan
in Australia and Asia

# CONTENTS

# PREFACE

My first memory of flyingfish (spelled as one word according to biologists) is from the fall of 1979 when I was sailing in the Mediterranean. While standing in line at a bank in Barcelona, during a month or so of wandering around Spain, a friend and I had met somebody looking for temporary crew for his motor yacht. It sounded like fun, and we spent several days sailing off the coast before entering Gibraltar (illegally, as we later found out) and then figuring how to get back into Spain, but that's another story. At sea in the lazy blue waters we saw a few seabirds, mainly shearwaters and gulls—but also flying fish (at that time I didn't know it was one word), amazing silvery creatures that shot out of the water and flew away from the yacht on stiff "wings." Wow, were those things cool! Time and again since then, I've vicariously experienced that initial wonder when I've seen people encounter their first flyingfish, usually with a gasp of surprise, whether off the coast of North Carolina or southern California or out in the tropical Pacific, home to a particularly high diversity of colorful species.

The Mediterranean was an appropriate place to make the acquaintance of flyingfish, for it is whence these animals were made known to science. The first flyingfish was named by Carolus Linnaeus in 1758, as *Exocoetus volitans*. Derived from the Greek, *Exocoetus* means "sleeping outside," which reflects the belief held by early Mediterranean sailors that flyingfish left the oceans at night to sleep on the shore.

One way or another, flyingfish feature in diverse aspects of human endeavor, from war to cuisine. In the early 20th century, the aerodynamics of flyingfish were studied by engineers in terms of airplane design, a flyingfish being a better scaled-down model for fixed-wing aircraft than a bird, with its actively flapping wings. The French word for flyingfish is *exocet*, and this name was given to the French-made, guided antiship missile that "flies" low over the water like a giant, deadly flyingfish. The flyingfish is the national animal of the Caribbean island of Barbados (sometimes called the "land of the flyingfish") and also the national dish: The fish are steamed and served with gravy and *cou-cou*, a mixture of cornmeal and okra. Half a world away, in Japan, some types of flyingfish and their eggs (known as *tobiko*) are used for sushi.

Despite being widespread in the world's oceans, flyingfish remain poorly known to most people. Yes, lots of fish can jump out of the water, but flyingfish have refined this to an art form—when they enter the air they can stay there long enough to be seen and enjoyed. You still need to be quick and in the right place, but it is possible to truly appreciate these remarkable animals when you really have a chance to see them—and with luck capture them in photos. I hope this book opens your eyes to another wonder of the oceans, the bodies of water that surround the land we live on and cover two-thirds of our planet. Note that the common names used here were created by field observers and in most cases can't be matched to formal scientific names (see "How Can I Identify Them?").

# THE AMAZING WORLD OF FLYINGFISH

Being able to sustain flight is all about balancing weight and wing area in combination with enough forward momentum. Whales, such as the Humpback Whale (opposite above right), are simply too heavy to leave the water for more than a breach, even with their long pectoral fins.

Rays (opposite below left) and dolphins (opposite below right, Pantropical Spotted Dolphin) can leap higher than a whale, but still don't have the right anatomy to sustain flight through the air.

# WHAT IS A FLYINGFISH?

Many types of marine animals leap out of the water, from dolphins, rays, and whales to sea lions, penguins, and squids, but flyingfish are in a league of their own. Flyingfish make up a specialized group of bony fishes placed by biologists in the family Exocoetidae, and are closely related to a few other fish families, including the needlefish (family Belonidae) and halfbeaks (family Hemiramphidae). Needlefish and halfbeaks, along with some other fish, can leap from the water but are not anatomically equipped to sustain their aerial travel beyond a few feet, even with some tail waggling to give them a bit of extra distance. The fossil record reveals that other, distantly related types of flyingfish (family Thoracopteridae) lived in the Middle Triassic period (some 240 million years ago), showing that the ability to fly has evolved at least twice in the world of fish.

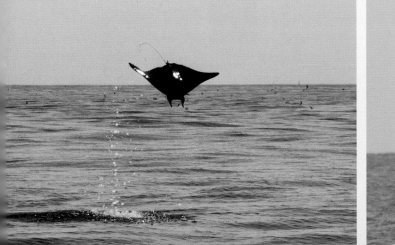

Many fish leap from the sea into the air, such as these baitfish off the coast of California (right, probably Pacific Saury), but their fins are simply too small to support them for any distance. They travel by "porpoising" like dolphins or at times by waggling their tails briefly to stay out of the water.

The closest relatives of the true flyingfish include the halfbeaks (below left, © 2011 J. Douglas Hanna), which can travel some distance over the water by beating their tail on the surface—but their small pectoral fins cannot support flight.

Among oceanic inhabitants that "fly," the closest to flyingfish may be flying squid (below right), which use their spread "tails" and flattened tentacles to create surfaces that enable lift. Flying squid can make short sailing flights ("tail" first) after shooting from the sea, presumably powered by the forceful water jets they squirt out.

Today's flyingfish are slender and streamlined, like torpedoes or cigars, and are characterized by very long pectoral fins (often referred to simply as "wings") and an unevenly forked tail, in which the lower fork is distinctly longer than the upper (in most fish the tail forks are about equal in length, or the upper fork is longer). Another adaptation of flyingfish is their hardened lower jaw, which protects the fine mouth bones from being smashed when hitting the water at high speeds.

When an adult flyingfish is swimming, its wings are normally held closed against the body and are always shorter than the body length. While the long pectoral fins may produce drag in the water, which might be a liability to swimming quickly, these fins redress any short-comings by transforming into wings that allow the fish to fly. The flyingfish's structure is a compromise, analogous (in reverse) to that exhibited by some seabirds with a reduced wing area that enables them to swim better underwater at the cost of labored flight, or, in the case of penguins, at the cost of becoming flightless. Recent videos of juvenile flyingfish have revealed them swimming with their forewings fully spread and raised and their hind wings spread and lowered, perhaps to appear larger or to disrupt and distract from the classic fish shape they would make with the fins closed.

Viewed side-on (top right) or without their wings spread, flyingfish often look quite unremarkable, like minnows or some other nonde-script silvery fish. But in the air the long pectoral fins act as wings and the lower tail fork as a propeller, and together they allow flyingfish to do what they do—and transform themselves from pedestrian into spectacular, as with this Atlantic Patchwing (middle right) and Bonin Windshield (bottom right).

# WHERE DO FLYINGFISH LIVE?

Living only in seas and oceans, not in freshwater, flyingfish can be found around the world in surface waters warmer than about 68° F (20° C), especially the blue waters of equatorial regions and the tropics. Although these oceanic waters contain many species of fish and other animals, they are relatively lifeless overall, the equivalent of deserts on land. There is little plant or animal life to intercept the light passing through them, and thus they look blue, like the sky.

While it has been suggested that the warm-water environment may be linked to helping cold-blooded fish achieve the speeds needed to fly, it may simply be that this is their ancestral home, as with other fish families restricted to tropical waters. Regardless, flyingfish are among the commonest fish in tropical surface waters, although their distribution tends to be patchy, as they track their shifting food resources over large areas of ocean. Some species inhabit inshore waters, but most live out in the open ocean. Within warm ocean environments, flyingfish comprise a key part of the ecosystem, forming a link between plankton and the larger predators, such as tunas and dolphins, and they are the most important prey item for many tropical seabirds.

Flyingfish occur singly or in schools of tens or even hundreds of individuals. The two-wing species often fly in single-species groups, and more often live in larger schools, whereas the four-wing species tend to be found in smaller groups, usually of five to 20 individuals, within which different species may associate.

Good places to see flyingfish include Hawaii and the Caribbean, as well as the warm Gulf Stream waters that bathe the U.S. East Coast from Florida north to the Carolinas. They can also be found farther north in warmer subtropical waters, as in summer and fall off southern California (as seen on boat trips out to the Channel Islands) and north to waters off New England and Nova Scotia.

# HOW MANY KINDS ARE THERE?

Scientists are uncertain how many different species of flyingfish are out there, and surely new species remain to be discovered. More than 150 types have been described over the years (juveniles often look so different from adults that they have been described as separate species), but modern fish biologists consider the total to be in the range of 60 to 70 species. There are approximately 32,000 species of fish worldwide, about 18,000 of which occur in the oceans; of these, therefore, fewer than half a percent are flyingfish. Despite being few in terms of species, however, flyingfish are among the most abundant fish in the surface waters of their open ocean habitat, which covers a large area of the planet.

The different species of modern flyingfish are grouped by scientists into seven genera (singular, *genus*), which are groups of species with shared and inherited characteristics. The flyingfish can also be viewed simply in terms of "two-wing" or "four-wing" species, a division based on the relative sizes of their "forewings" (pectoral fins) and "hind wings" (pelvic fins). Just as in birds, the size and shape of the wings of flyingfish affect how well and how far different species can glide. It is thought that the four-wing mode of flight evolved from the two-wing mode, and the former thus represents the most highly evolved aerial flight in fish.

The seven or so species of two-wing flyingfish (genera *Exocoetus* and *Fodiator*) are relatively small, usually up to 6–7 inches (15–18 cm) in length, and their enlarged forewings comprise most of the lifting surface. Their hind wings are variable in size and rarely visible in

Some small flyingfish, such as the Pixellated Midget (above), may simply be juveniles that change appreciably in shape and even in pattern as they grow.

Although some of the largest types, such as the spectacular Black-eyed Blushwing (above), appear distinctive, many flyingfish species closely resemble one another and can be difficult to distinguish in the laboratory, let alone at sea.

flight; even when spread they may be concealed by the large fore-wings. Two-wing species such as the Small Clearwing (top right) typically glide relatively short distances but can still cover 50 feet (15 m) or more in a glide. These species often make only a single glide before splashing back into the sea.

The three species of sailfin flyingfish (genus *Parexocoetus*) are usually considered as two-wing types, although they have somewhat enlarged hind wings. In addition, their large, sail-like dorsal (back) fin can flip sideways and in that position appears to provide a fifth "wing" for extra lift, as on the Oddspot Midget (opposite page; the dorsal fin has a big black spot). Sailfins often use the tail to power a second or third glide before they reenter the water.

In the 50 or so species of four-wing flyingfish (genera *Hirundich-thys*, *Prognichthys*, *Cheilopogon*, and *Cypselurus*), the forewings of most species are longer and relatively narrower than those of two-wing species (and thus have a better lift-to-drag ratio), while the hind wings are relatively broad but not as long as the forewings (as in the Leopardwing, bottom right). This wing structure helps four-wing species glide for longer distances than two-wing species. These are the flyingfish that most people notice, for they are relatively large and stay airborne long enough to be seen, often using the tail to spur multiple glides in one flying episode.

# HOW BIG ARE THEY?

Some flyingfish reach 20 inches (50 cm) in length and about 30 inches (75 cm) in "wingspan," but adults of most species are 6–12 inches (15–30 cm) long and have wingspans of 9–18 inches (23–46 cm). Some flyingfish lay sticky eggs on seaweeds (such as the golden Sargasso weeds that characterize the Gulf Stream), palm fronds, pieces of wood, and other floating debris, whereas other species, typically those found farther offshore, lay buoyant eggs that float near the sea surface. Bathed in warm waters, the eggs usually hatch within a week or so, and young flyingfish grow quickly, usually attaining their distinctive adult shape and size within only one to two years. As adults, the largest species weigh up to about 1.5 pounds (680 g).

While some fish biologists may fantasize about flyingfish similar in size to trophy sport fish (right, a computer-enhanced image of *Hirundichthys speculiger*, © Robert L. Pitman), in real life these are fairly small fish, rarely exceeding about a foot (30 cm) in length, as shown above.

Many of the flyingfish you may see are likely to be young, not the full-size adults portrayed in field guides. The smallest young that fly, sometimes known as "smurfs" (three are pictured at right), are only up to about an inch (2.5 cm) across; as they leave the sea they often look like small silvery bubbles or disks that flip out for only short distances, seemingly at the mercy of the wind. In some flyingfish, perhaps especially the four-wing species, the color patterns of smurfs and other young juveniles are quite distinct from those of a full-grown adult. In other instances, smurfs appear to be recognizably similar to an adult.

Of the three smurfs at right, it is difficult to imagine what the upper two might become, but the one in the bottom image looks as if it may develop into an Oddspot Midget (see p. 9). In some four-wing species, older juveniles such as the Sargassum Midget (below, which was about 2 inches or 5 cm across) have intricate patterns that serve as camouflage amid the patches of Sargasso weed they inhabit. One juvenile similar to this was examined genetically and proved to be a young Atlantic Necromancer (see p. 31).

# HOW DO THEY FLY?

Two obvious and linked questions are: How do flyingfish fly, and do they really fly? Looking down from the bow of a boat into pellucid water, you can sometimes see flyingfish below the surface, twisting, turning, and darting quickly before they shoot abruptly from the water and zip away through the air with remarkable speed and grace. To generate the initial thrust to power flight it appears that a fish bends its body sideways to nearly 90 degrees and then "snaps" back into a straightened shape, as do hunting pikes or barracudas to generate a short-term but powerful thrust. This initial thrust is enhanced by tail movement, which continues as a fish breaks the surface and uses its tail as a propeller, whipping it quickly from side to side to achieve maximum speeds of over 40 mph (65 km/h). As a rule, when a vessel is approaching, flyingfish fly only if the boat is moving faster than they can swim: A boat moving at 5 knots tends not to flush them, but at 10 knots the fish usually take to the air.

Sometimes, especially among the two-wing species, a fish appears to generate the full power needed for a short flight before breaking the surface and thus makes a clean break into the air. In windy conditions, emerging flyingfish can be swept into a steep loop some 50 feet (15 m) or more above the sea surface before splashing back down near the point they left the water—and they can even get deposited high and dry on the decks of boats and ships.

More often, at least among the four-wing species, the tail continues beating from side to side as the fish moves from water into air, often emerging at a fairly steep angle, which presumably helps reduce drag. The elongated lower tail fork thrashes and splashes in the surface water until the fish achieves takeoff speed; hence the zigzag patterns observers see on calm water as flyingfish flee from a boat. The forewings are held at least partially spread as the tail beats the sea surface, but the hind wings, which act as control and elevating surfaces, usually open only when the fish loses contact with the water. The dorsal fin may be held raised while the tail is beating, but often it is folded, and while the fish is gliding both the dorsal and anal fins are often kept folded, perhaps to reduce drag.

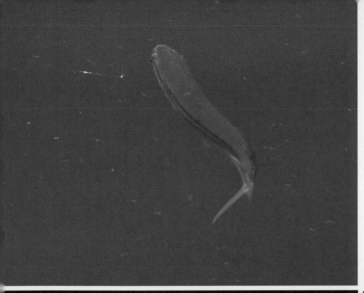

These three images, taken within a second, show an Atlantic Patchwing making the remarkably rapid transition from a streamlined, swimming "torpedo" into a thrashing, taxiing, flying machine. Within another fraction of a second it was airborne and well away from our boat.

As a fish leaves the water, well-developed muscles pull its forewings forward, then spread and lock them open for flight. Reentry into the water can be a fairly graceful slide, with the tail entering first, or simply a plop and splash; if a booby is closing in, the fish can fold its wings and drop back quickly into the sea.

With increasing body size, the wing area of a flyingfish supports a proportionately greater weight. Consequently, the larger species of flyingfish need to move faster in order to fly, but once airborne they can travel farther than smaller species. Unlike birds and bats, flyingfish do not beat their wings, and once airborne they are simply gliding on fixed wings, typically following a low, arcing trajectory within a few feet of the sea surface. Thus, they might more correctly be named "glidingfish" than flyingfish.

In addition to being powered by the tail, prolonged flight is aided by updrafts of wind blowing over the sea surface and by an aerodynamic process known as "ground effect," which occurs as swirls of air generated by the wingtip hit the water and effectively create a cushion of air that helps keep the fish aloft. (Pelicans employ a similar effect when sailing low over the waves.) Partly due to ground effect, flyingfish have a much more level flight path than other gliding creatures, such as flying squirrels and flying frogs.

On this Big Raspberry (right), the dorsal fin (with a big black spot) is held raised in the taxiing stage (top) but is then folded down when the fish takes flight (bottom), at which point the hindwings are deployed to help keep the fish airborne longer.

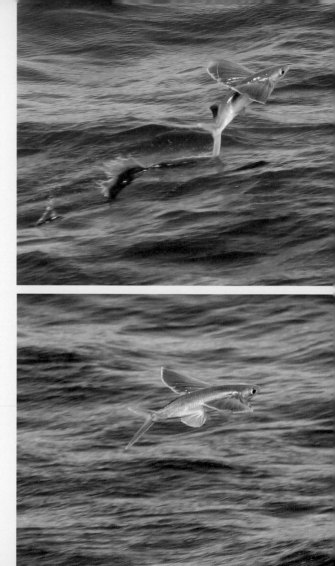

In taxiing mode, a flyingfish (Atlantic Patchwing top row, Leopardwing bottom row) often maintains a fairly upright pose, perhaps carried over from its steep angle of exit from the water. The forewings are held at least partially spread as the tail beats the sea surface, and when the fish has reached take-off speed its hindwings are often deployed, which may serve to lift the tail end of the fish into the more horizontally oriented plane typical of flight mode, allowing it to sail low over the water and take advantage of "ground effect."

If a flyingfish doesn't want to wait until its glide comes to an end to reenter the water, all it has to do is fold in its wings, which causes it to lose lift and plop back into the ocean, as shown by this Sergeant Pepper (left) in the western tropical Pacific.

At other times a flyingfish may simply slide back into the water, as shown by the Atlantic Patchwing (above, top) and Rosy-veined Clearwing (above, bottom). On reentering the water, the fish can quickly fold its wings closed against the sides of the body.

Glides made by flyingfish are typically in the range of 50–300 feet (15–90 m), and are sometimes much shorter. However, some species, in particular the four-wing models, can kick off again with a burst of tail motion rather than dropping back into the ocean. By folding the hind wings, the four-wing species can even induce the tail to dip back into the water and repower or stabilize their flight as needed. In this way, flyingfish can travel for considerable distances, mainly at speeds of 20–40 mph (32–64 km/h) and sometimes execute ten or more separate glides in a single flying event. The longest flights involve multiple glides and can cover a quarter mile (0.6 km) or more in total distance.

While most flights are of only a few seconds, the four-wing species in particular often stay airborne for 10–30 seconds, sometimes longer. The longest flight recorded was in 2008 off Japan, when a flyingfish was filmed in the air flying alongside a ferry for 45 seconds.

In flight, the forewings and hind wings can be held in different planes, as on this Yellow Bandwing (left), presumably to help with stabilization and perhaps to allow slight changes in direction.

# WHY DO THEY FLY?

So fish can fly (or at least glide), but why do they? At one time it was suggested that the flight of flyingfish offered a more efficient means of travel, as it does for penguins, dolphins, and sea lions, which leap from the water in "porpoising" mode when they need to travel their fastest. While this explanation may hold for the schools of small bait-fish sometimes seen porpoising (p. 2), studies of physiology do not support this theory for flyingfish. And as anyone watching flyingfish flee from the bow of a ship or boat can attest, flight is certainly used to escape from potential predators. The same is true for flying squid, which are found worldwide and usually seen singly or in groups shooting away from oncoming vessels (p. 2). Unlike flyingfish, flying squid usually shoot out only once and are unable to sustain their movement through the air, as flyingfish can do by using their tail.

Several marine mammals, such as Pacific White-sided and Northern Rightwhale dolphins (above) and New Zealand Fur Seals (below left), along with penguins (below right, Gentoo Penguins), can travel more quickly by "porpoising," because the air offers less resistance than the denser water. But for flyingfish becoming airborne appears to be a mode of escape, not a form of economic travel.

The consensus today is that flyingfish have developed the ability to fly in order to escape from predators. The warm surface waters are home to any number of fast-swimming predators, and seen from underwater the sea surface appears partly like a mirror, making it difficult for underwater predators to see or follow fish in the air. Some fast-swimming predatory fish might be able to catch up with flyingfish when they reenter the water, especially during the daytime. Other predators, however, are nocturnal and may have difficulty seeing and following airborne flyingfish at night. It has been suggested that flight in fossil fish independently evolved in response to predators such as marine reptiles.

As is typical of many smaller fish, flyingfish mostly eat plankton, the tiny animal and plant organisms floating in the sea. To "hunt" for plankton doesn't require any great speed, but flyingfish themselves are edible, and they live in the open ocean, where survival of the fastest is the maxim. This is the home of fast-swimming predatory fish such as tuna, dolphinfish, and Swordfish, as well as of dolphins and some fast-flying seabirds—all providing an impetus for the evolution of flyingfish in such areas. Think of antelopes grazing on the African savanna—hunting grass isn't hard, but avoiding lions is.

I remember one day in April 2005 going ashore before dawn to see turtles lay their eggs at Ascension Island, in the equatorial Atlantic Ocean. Between the well-lit ship and the dock our Zodiacs were a perfect stage from which to see Bottlenose Dolphins chasing flyingfish. The elegant, playful, bow-riding dolphins of the daytime were changed, like Jekyll and Hyde, into lightning-fast, frenzied killing machines that slashed the water into phosphorescent streaks as they drove flyingfish this way and that, sometimes straight into our Zodiacs!

Dolphinfish or mahi-mahi (above), also known as dorado for its intense golden coloration, is a popular sportfishing target. It is one of many fast-swimming predators that eat flyingfish; this individual had several whole flyingfish in its stomach when we cleaned it for dinner.

The Gulf Stream is often marked by lines of golden Sargasso weed (below) where different water masses meet. Plankton and debris are often concentrated at these weed lines, which are also good places to see flyingfish.

It has been suggested that two-wing and four-wing species have evolved in response to different predators. Some two-wing species are hunted by schools of fast-swimming tuna, which do not pursue the flyingfish once they leave the water; the two-wing design enables quick exit of the water, without the need to use the tail for extra lift, and shorter flights suffice to avoid the tuna. Conversely, some four-wing species are hunted by dolphinfish, so named because they leap from the water like dolphins, perhaps a specific adaptation to following flyingfish, their main prey. To avoid such predators requires the counter-adaptation of longer and faster flights that can involve changes of direction.

Leaving the water, however, can bring its own perils, for airborne flyingfish are vulnerable to being snapped up as food by agile seabirds such as boobies, frigatebirds, gadfly petrels, and terns. Boobies have even learned that ships will flush flyingfish, and sailors in tropical waters often get to watch boobies accompany their ships, riding updrafts beside the bridge or the bow for hours at a time. On seeing a flyingfish the boobies tilt and dive, like fighter planes, swooping down over the water in an attempt to catch the hapless fish. While standing on the bow photographing flyingfish, I have noticed that three different booby species (Red-footed, Brown, and Masked) have different hunting styles. I've also seen that boobies can detect the fish several seconds before they break the surface—giving them an unfair advantage over the photographer!

The Red-footed Booby (below, and preceding page) is the most lightly built and aerodynamic species, and it can often snatch up a fish in the air. The Masked Booby is more heavily built and less maneuverable, and by the time it gets there the fish is usually back in the water. Its technique is then to arrive in time to plunge dive at a steep angle and grab the fish underwater. Between these extremes lies the Brown Booby (opposite page), which is slower and heavier than the Red-footed but lighter and quicker than the Masked. Brown Boobies sometimes get the fish before it hits the water, but often they arrive just as the fish is reentering and then follow it in with a shallow-angle dive that can force the fish back into the air; the booby then snatches the fish on the second, unintentional flight.

Unlike the more agile Red-footed Booby, Brown Boobies often arrive a second too late to snatch a flyingfish in the air—so they chase it underwater with a shallow-angle plunge dive. Sometimes the fish shoots back out into the air, only to be snatched on the rebound.

Flyingfish are also eaten by people. They may even have helped early Polynesian sailors survive on the long, oceangoing odysseys by which they colonized remote Pacific islands. On his famous *Kon-Tiki* expedition from Peru to Polynesia, the Norwegian explorer Thor Heyerdahl recounted that the cook's first duty each day was to collect the flyingfish that had landed on the raft at night, and fry them up for breakfast. Today, annual catches worldwide number in thousands of tons, attesting to the abundance of flyingfish and to their potential importance in some tropical areas as a food source. Commercially, they are caught mainly by gillnetting: Curtains of net are hung vertically in the water column and the fish swim into them. In other areas, such as the Solomon Islands, they are attracted at night by lights and caught by nets held in the air. Flyingfish eggs (known as *tobiko*) are a popular feature of Japanese cuisine, as in sushi (right).

Inevitably, given human greed and our unrealistic population levels, flyingfish have been subject in some areas to overfishing, and the resulting scarcity has even led to international conflict. In 2006, the council of the United Nations Convention on the Law of the Sea fixed the maritime boundaries between Barbados and the Republic of Trinidad and Tobago, following tensions stemming from a flyingfish dispute. Flyingfish migrate annually from coastal areas of South America north to waters around Barbados. In recent years, numbers have been decreasing and migrations have not extended as far north, likely the result of overfishing, perhaps in conjunction with marine pollution. The ruling stated that both countries must preserve stocks for the future. However, without monitoring and proactive management, this seems no more likely to happen than in similar situations anywhere else in the world, where fish species after fish species has been overfished and pushed to or beyond the edge of sustainability.

# WHAT COLORS ARE THEY?

The bodies of flyingfish are mostly a generic silvery color, with a dark blue topside and often with a white belly. This countershading helps protect them from predators—when they are seen from above the dark back blends with dark ocean depths, and from below the pale belly blends with light coming from above the surface. Some species may undergo color changes that relate to courtship, when parts of the body flush reddish, but details of these changes are not well known. The wings, however, are a different matter, at least among the four-wing species.

All two-wing and many four-wing flyingfish have dirty pale grayish or mostly clear wings, but the spread wings of some four-wing species reveal beautiful pastel pinks, blues, purples, and lemon yellows, often offset with black pigmentation. Some of the pink coloration may come from blood vessels, which also help to stiffen the wings. Nearly all these bright colors are "fugitive," meaning they fade or disappear within minutes of death, and in some cases the colors may be simply artifacts of refractive sunlight created as fish fly through the air, as with the colors of the sky we see under different lighting and atmospheric conditions. Moreover, because a number of fish can see into the ultraviolet spectrum, the wings of some flyingfish may also include colors we cannot see.

Two of the more attractive flyingfish include the Purple Haze (top right) and the Yellow Bandwing (bottom right).

Pink is one of the most commonly seen colors on flyingfish. Clockwise from top left: Rosy-veined Clearwing, Big Pinkwing, Thrushwing, Apache Pinkwing. Opposite page: Freckled Pinkwing.

Purples and blues (Violaceous Rainmaker, top left; Solomon Cerulean, top right) are less common than pinks, and many of the larger flyingfish types have fairly clear, plain wings (Large Clearwing, bottom left). Bold and complex blackish patterning is seen mainly on smaller flyingfish (Double Hyena, bottom right), perhaps juveniles of species that grow up to have clear wings.

A few species have largely blackish wings (Pacific Necromancer, top left; Atlantic Necromancer, top right), and some have striking yellow coloration (Yellow Bandwing, bottom left; Leopardwing, bottom right).

Why some flyingfish should have such colorful wings seems to have evaded satisfactory explanation. The complex patterns on the spread wings of many juveniles, such as those on the Sargassum Midget (opposite page) act as camouflage when the fish are swimming among weed masses. But why would some adult flyingfish have such bright and bold colors and patterns?

It may be that flyingfish flash their wings underwater to startle and confuse predators, as do some other (nonflying) fish with large, colorful fins. Or perhaps the wings are flashed in underwater courtship display. Or perhaps the colors help different species recognize each other. For now, speculation exceeds study of this subject.

# HOW CAN I IDENTIFY THEM?

A big hurdle to identifying flyingfish in the field, in flight, is that scientific descriptions are usually based on specimens examined in museums, and involve counts and measurements such as fin lengths, the number of rays (bony supports) in a fin and whether or not the rays are divided, the numbers of scales on different parts of the body, and so on. For example, an entry from a formal key to flyingfish reads: "Anal-fin rays 10 to 12; longest dorsal-fin rays scarcely reaching origin of upper caudal-fin lobe; predorsal scales 16 to 21" (Parin 1999). While useful for scientists with a fish in hand, such descriptions are all but impossible to match with even the best at-sea photos. Moreover, while technical identification keys sometimes refer to black, pigmented areas on the wings and tail, they rarely mention the stunning rainbow colors often seen on live fish. This is because museum specimens tend to be preserved with the wings closed, and, as well as fading naturally, the bright colors may be leached out by chemicals used to preserve the fish.

Matching an image of a flyingfish aloft, like the one above right, to a formal identification key based on museum specimens is all but impossible. Even determining which genus it might belong to is not straightforward. For now, flyingfish-watchers at sea call this a Fenestrated Naffwing.

The pale median band on the wings of the flyingfish at right (presumed to be a Blue Bandwing) was likely quite bright blue in life, but it has faded appreciably within a few hours of death. A specimen pickled in a museum jar would likely lose a lot more color.

Popular field guides to marine fishes often include flyingfish among the species they treat, and they provide English names for them. These guides, however, are designed basically for in-hand identification, and the illustrations are typically based upon dead specimens, which are shaped very differently from a fish in flight.

Thus, field guides typically portray the wing base as narrow, which it is when a flyingfish is swimming. However, a fish with a wing shaped that way in the air would be unable to glide any distance. In life the wings fan out and merge along the sides of the body to create an aerodynamic seal that allows lift and flight.

From the descriptions and portrayals of flyingfish in books and scientific keys it can even be difficult to determine if photos of a fish in flight refer to a two-wing or a four-wing species. In general, the two-wing species have relatively short and broad wings, as shown in the Delta Cenizo (following page, bottom right) and the Small Clearwing on p. 8. The four-wing species tend to have relatively narrower but longer forewings (as seen on the Solomon Cerulean, p. 30, and Atlantic Necromancer, p. 31), and their hind wings may be withdrawn in flight.

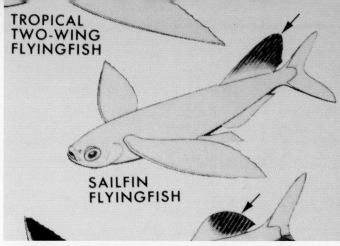

TROPICAL TWO-WING FLYINGFISH

SAILFIN FLYINGFISH

Traditional field guides do not show flyingfish as they truly appear when flying. Compare the shapes and positions of the fins, and even the tail pattern, between the field guide sketch of a sailfin flyingfish (top right) and the in-flight Oddspot Midget (bottom right)—yet apparently these are the same species, or at least in the same genus, *Parexocoetus*.

There remains an unfilled niche for a field guide that portrays flyingfish as observers see them in the air, although for many years researchers studying seabirds and marine mammals have known about and recognized different types of flyingfish—for example, those known as "pinkwings," "blackwings," and "bandwings." A provisional, at-sea photo key was developed in 2008 on the second Western Pacific Odyssey (WPO) cruise from New Zealand to Japan aboard the vessel *Spirit of Enderby*. These trips are run by the New Zealand company Heritage Expeditions, mainly for bird-watchers. But birds tend to be few in the blue equatorial waters (remember, it's a desert, even though it's full of water), and attention sooner or later shifts to flyingfish.

On the inaugural WPO cruise, in 2007, I had taken many photos of flyingfish and learned something of the trials and tribulations involved. In 2008, together with fellow enthusiasts Rob Tizard (a U.S.–born conservationist working in Southeast Asia), John Ryan (a British doctor and lifelong birder), and Michael Boswell (a keen birder and photographer from western Canada), I started to develop a catalog and key for the flyingfish we saw on the WPO route. By the end of the trip we had identified and named 51 "types" of flyingfish, mapped their distribution by latitude, and made an informal, photo-identification guide. Some of these 51 types surely represent different ages and perhaps even sexes of the same species, but even so, we could identify them at sea and give them names.

The wing base of a flyingfish, which appears narrow in a dead specimen (left and below, *Hirundichthys rondeletii*, found off Cape Cod), is transformed in flight, spreading out to form an aerodynamic seal along the sides of the body so the fish can stay in the air (bottom, Delta Cenizo, *Exocoetus* species).

In naming the fish, we soon gained an appreciation for the challenge of coming up with lots of new names. After obvious names like Large Clearwing, Small Clearwing, Yellow Bandwing, and so on are taken, where do you go? Fortunately, my colleagues were up to the task (John was well schooled in British moths, some of which have rather creative names, to say the least), and sometimes we farmed the job out to other interested passengers ("Hey Linda, could you take these pictures and think of a name?"). A printout of this key sits in the library on board the *Spirit of Enderby*, and passengers can now put names to many of the flyingfish they see or photograph in the western Pacific. This key can also be viewed online (and downloaded) at: http://www.offshorewildlife.com/OffshoreWildlife/Flyingfish_of_the_WPO.html.

When we field-tested the identification key on the WPO in 2011, we found only five new types to add to the list. What was even more interesting, we found most of the named types again and in about the same places as in previous years. Not surprisingly, these flyingfish—just like birds or any other organisms—have their own specific, somewhat predictable seasonal distributions and habitats within the wider realm of the tropical ocean.

In coming years I hope a greater awareness of flyingfish will develop, and more at-sea identification keys will be compiled for different parts of the oceans. One day, presumably, we'll be able to link the informal names we created to the scientific names, but for now the hobby of flyingfish-watching is in its infancy—and it sure is fun to be a kid again! Moreover, when humans start to recognize and name things, we have entered the first stage of understanding, and with understanding will come more knowledge, which may help us to protect the oceans, the cradle of life on Earth.

Even a flyingfish as "unmistakable" as this large Solomon Cerulean (above) can show variation in how far the hind wings can be spread out or retracted (these images of the same individual were taken less than a second apart). This might be confusing if you are trying to tell a four-wing from a two-wing species.

# A NOTE ON THE PHOTOS

All but two of the images in this book were taken by me, most with a Canon 20D camera body (such as the Purple Haze, top right), some with a 40D, and lately a 7D (Sargassum Midget, bottom right), mainly using a 100–400mm zoom lens. Other than the "trophy catch" image (p. 10), none of the images has been manipulated or altered beyond simply cropping, plus tweaking the exposure, sharpness, and color balance. To obtain these photos, I spent several hundred hours perched at the bow of boats and ships, sweating in the baking tropical heat, staring at the mesmerizing clear blue waters, and only once suffering mild heat stroke. I also obtained a few thousand fuzzy images, or images of blank water or water punctuated by only a splash as a fish escaped back into its watery home. Most images came from the western tropical Pacific Ocean, shot during three cruises between New Zealand and Japan (aboard *Spirit of Enderby*); a good number came from Gulf Stream waters, mainly off Cape Hatteras, North Carolina (aboard *Stormy Petrel 2*, captained by J. Brian Patteson); and one from the warm waters off San Diego, southern California.

# ACKNOWLEDGMENTS

For helping me see and photograph flyingfish, for discussions about them, and for help naming them and gathering literature, I thank Heritage Expeditions, WINGS, Rob Tizard, John Ryan, Michael Boswell, Doug Hanna, Tom Blackman, Linda Utterberg, J. Brian Patteson, Kate Sutherland, Dave Shoch, Robert L. Pitman, Seabird McKeon, Burr Heneman, Broni Alberti, Tim Howell, Terry Hunefeld, Bill Bennett, and Megan Elrod. Doug Hanna kindly contributed his photo of a halfbeak (p. 2); Robert L. Pitman and Cornelia Oedekoven produced the "giant flyingfish" trophy catch image (p. 10); and Houghton Mifflin Harcourt Publishing Company gave permission to reproduce part of a plate from *A Field Guide to Atlantic Fishes of North America* (p. 35), by C. R. Robins & G. C. Ray (© 1986). I am indebted to Pitman, McKeon, Heneman, and two anonymous reviewers for their comments on the manuscript; any errors remaining are my responsibility.

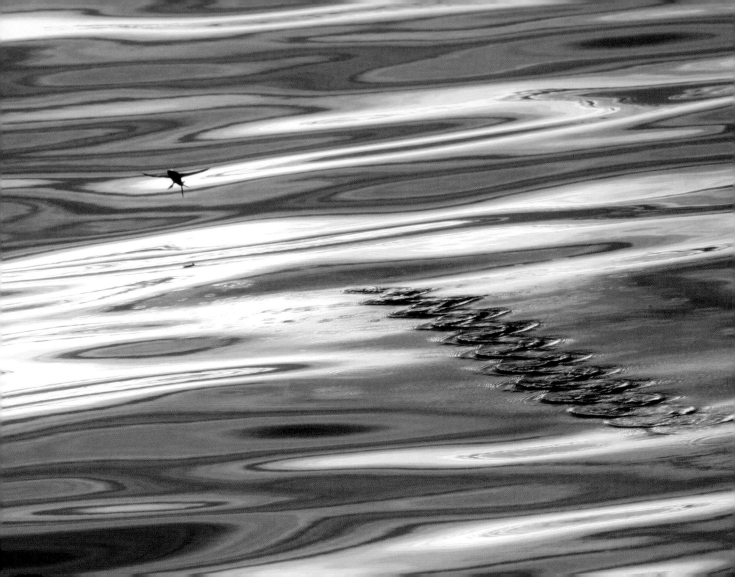

# REFERENCES

Breder, C. M. Jr. 1930. "On the structural specialization of flying fishes from the standpoint of aerodynamics." *Copeia* 4:114–121.

Davenport, J. 1994. "How and why do flying fish fly?" *Reviews in Fish Biology and Fisheries* 4:184–214.

Fish, F. E. 1990. "Wing design and scaling of flyingfish with regard to flight performance." *Journal of Zoology, London* 221:391–403.

Heyerdahl, T. 1950. *Kon-Tiki: Across the Pacific by Raft*. Rand McNally & Co., Chicago.

Hubbs, C. L. 1918. "The flight of the California flying fish." *Copeia* 1918 85–88.

Hubbs, C. L. 1933. "Observations on the flight of fishes, with a statistical study of the flight of the Cypselurinae and remarks on the evolution of the flight of fishes." *Papers of the Michigan Academy of Sciences* 17: 575–611.

Jacobs, G. H. 1992. "Ultraviolet vision in vertebrates." *American Zoologist* 32(4):544–554.

Kutschera, U. 2005. "Predator-driven macroevolution in flyingfishes inferred from behavioural studies: historical controversies and a hypothesis." *Annals of the History and Philosophy of Biology* 10:59–77.

Lewallen, E. A., R. L. Pitman, S. L. Kjartanson, & N. R. Lovejoy. 2011. "Molecular systematics of flyingfishes (Teleostei: Exocoetidae): evolution in the epipelagic zone." *Biological Journal of the Linnaean Society* 102:161–174.

Parin, N. V. 1999. "Exocoetidae. Flyingfishes." In *FAO Species Identification Guide for Fishery Purposes: The Living Marine Resources of the West Central Pacific*, vol. 4, *Bony Fishes*, part 2 (Mugilidae to Carangidae), eds. K. E. Carpenter & V. H. Niem, 2162–2179. FAO, Rome.

Roberts, C. 2007. *The Unnatural History of the Sea: The Past and Future of Humanity and Fishing*. Octopus Publishing, London.

Robins, C. R., & G. C. Ray. 1986. *Peterson Field Guide to Atlantic Coast Fishes*. Houghton Mifflin Co., Boston.

Xu, G.-H., L.-J. Zhao, K.-Q. Gao, & F.-X. Wu. 2012. "A new stem-neopterygian fish from the Middle Triassic of China shows the earliest over-water gliding strategy of the vertebrates." *Proceedings of the Royal Society B* 280:2012.2261. doi.org/10.1098/rspb.2012.2261.

# INDEX